MICHIGAN
ROSS SCHOOL OF BUSINESS

William Davidson Institute
AT THE UNIVERSITY OF MICHIGAN

case 1-428-925
February 19, 2010

Madras Cement

Case Summary

Madras Cements is a leading cement producer in southern India. The commodity nature of the industry makes cost efficiency the most important aspect of competition in the industry. Madras Cements was having problems controlling costs because of inefficient and insufficient reporting processes. The first attempt to improve the reporting process through enterprise resource planning (ERP) did not make any headway. The Madras Cements management applied the lessons learned from the failed ERP implementation effort to a second ERP implementation initiative and reaped substantial benefits. The case illustrates the process of successfully championing ERP implementation in a complex diversified conglomerate setting in a developing country. The case, particularly, highlights the significance of the social transformation realized by Madras Cements' team while implementing the enterprise solution after the initial unsuccessful attempt. Important aspects of ERP implementation like process redesign, change management, and importance of top management support are discussed.

A.V. Dharmakrishnan, CFO of Madras Cements, was quite pleased with how the day was progressing. It was the end of the quarter and, as the CFO of the company, it was his responsibility to close the books. A task that used to take days in just getting all relevant data from the different manufacturing plants and marketing locations that Madras Cements operated across southern India now was completed in real-time. This was a direct benefit of the IT-enabled business transformation he had championed.

Background

Madras Cements Ltd was the flagship company of Ramco Group that was founded by P A C Ramasamy Raja in 1938. The Ramco group of companies had multiple lines of business including textiles, cement, fiber cement products, software, surgical cotton, and bio-tech. The Ramco Group grew into a massive organization with annual revenue of $800 million, achieving international recognition for its quality products and services under the leadership of P R Ramasubrahmaneya Rajha. (**Exhibit 1** details the structure of Ramco group.)

GlobaLens℠

Published by GlobaLens, a division of The William Davidson Institute at The University of Michigan.

Sanjeev Kumar. Balaji Sudarsanan, and Adarsh Prabhakar prepared this case under the supervision of Professor M. S. Krishnan and Professor C. K. Prahalad. Rajeev Ved conducted the field study and initial documentation for the case. Ross School of Business, University of Michigan.

Madras Cements was started in 1957 with its first cement manufacturing plant at R R Nagar in the southern Indian state of Tamil Nadu. By 1962 the plant could manufacture 200 metric tons per day. The company expanded to include seven manufacturing facilities with a combined capacity of 11 million metric tons per year. In addition to cement manufacturing, the company operated several wind power sites with combined capacity of 185 MW that provided Madras Cements with a cheaper power supply. Within the Ramco group lines of business -- including textiles, cement, fiber, software, surgical equipment, and bio-tech -- the cement manufacturing operation under the Madras Cements banner remained the flagship company. The Ramco brand enjoyed a strong recognition in southern India and was strongly associated with quality. The company manufactured cement at five cement plants and two grinding plants in southern India. (**Exhibit 2** lists the manufacturing facilities of Madras Cements.)

Demand for cement is closely related to overall growth in the economy and particularly growth in the construction industry. Starting in 1997, the Indian economy had seen a significant increase in infrastructure-related projects. As a result the cement industry grew by a strong 8% annually since 1997. The cement industry in India was highly fragmented with over 50 cement producers and more than 140 manufacturing plants. During 1997-98 to 2008-09, the installed capacity of the industry increased at a CAGR of over 8% to 217.80 MTPA. In 2008-09, cement production from large units touched 181 million metric tons, indicating a capacity utilization of around 83%. Cement production depends on availability of limestone, which is heavy and is expensive to transport long distances. Hence, the industry was highly regionalized and cement units were concentrated in clusters close to limestone deposits. In India, as a result, 83% of domestic production took place in the seven states of Madhya Pradesh, Andhra Pradesh, Rajasthan, Gujarat, Karnataka, Tamil Nadu, and Maharashtra. Competition was regionalized because the low value to weight ratio of cement makes transportation over long distances uncompetitive.

The cement manufacturing business consumes vast amounts of power and is significantly affected by electricity prices. Coal and power account for 34% of the total cement production costs. High distribution costs are also a significant cost component and freight costs account for approximately 22% of total costs. These three are the most significant cost components in cement manufacturing. (**Exhibit 3** describes the process of converting limestone into cement.)

Cement is a commodity, and it was traditionally believed that there were very few opportunities for deriving premium prices. Every cement-manufacturing firm in southern India was subject to fluctuation in cement prices. (**Exhibit 4** presents historical cement prices in Madras Cements' key markets.) Because, traditionally, cement companies do not differentiate on quality, they need to ensure that their cost structure is lower than the industry average to be competitive and profitable. (**Exhibit 5** shows the trend in cement demand in India and **Exhibit 6** provides market share information of leading producers of cement in India.)

Achieving Operational Efficiency

The top managers of Madras Cements were becoming increasingly aware that in order to gain a competitive advantage in cement manufacturing they would have to ensure that their cost structure was low. Managing costs required a new level of transparency across the entire company. Starting with the first plant at R R Nagar, Madras Cements invested heavily in deploying an effective IT infrastructure at each of the plants in order to help monitor and control costs in a systematic function. With the active and enthusiastic support and guidance provided by the CEO, Venketrama Raja, the company launched many IT-related projects as early as 1983. He encouraged the management team to develop effective IT strategies that provided a competitive edge for the company.

Previous IT Initiatives

Before the 1980s cement was a controlled commodity. But the subsequent rise of the liberalized era led to an open market and increased competition. The Madras Cements management was clear that it could not yield to market forces, even as the cement industry had become highly competitive. Madras Cements knew that it could not do much about product prices in a fiercely competitive market. Thus, to be profitable, the need was to be a cost leader. This meant that it achieve cost effectiveness and consistent quality, apart from ensuring prompt response to customer needs. Accurate, transparent, and timely information was the key to achieve such operational efficiencies.

Moreover, in the '80s, due to price controls and a permit system for cement, Madras Cements had to deal directly with thousands of cement customers who wanted only 5 tons or 10 tons. The distribution system and the underlying accounting system became very complex, which called for extensive computerization. Madras Cements was perhaps the first in the cement industry to invest heavily in developing IT infrastructure. As early as 1983 it set up UNIX-based Mini running COBAL. That marked the beginning of the IT revolution in the company.

So the option was clear—that is, to adopt technology to beat operational complexities. And Madras Cements did just that! The opportunity for the company to profit in this era existed only by serving quality products to the customer at economical prices. The computerization helped Madras Cements to set up an efficient accounting system to deal with complexities of distributing cement, which was the need of the hour.

Further, to match global standards, Madras Cements adopted state-of-the-art technologies in its manufacturing units, deploying the fuzzy logic software system for process controls; programmable logic controllers; vertical mills for cement grinding; advanced X-ray technology for ensuring quality control, and so on. Once the process of adopting IT started in the '80s, Madras Cements didn't let the process slow down at any point and kept evolving its technological prowess. It later arrived at the point where almost all of the company's critical business processes were running on IT solutions.

Madras Cements was expanding very fast. From a mere 0.4 million tons per year capacity in 1980, it became an 11-million-ton-per-year company. From just a single plant in 1980, Madras Cements became a multi-unit company with five ultra-modern cement plants spread across three states.

The company's previous information management systems were not able to catch up with the multi-functional/multi-factory setup. Coordination between departments and people was becoming problematic and complex. This was the pre-ERP (enterprise resource planning) era when no real-time information was available. Besides, information was neither consistent nor transparent. There was no standardization of processes or transparency of data. People used their own formula and formats for analyzing and interpreting data and generating reports.

Madras Cements wanted to bring everyone onto one information platform by achieving transparency of data across units and departments. It wanted to standardize the way people looked at data, understood it, and interpreted it, so that they spoke the same language as far as data was concerned. The company's previous systems were not able to manage these growing complexities.

There was a dedicated IT support staff at the R R Nagar facility to support the growing IT needs of that manufacturing plant. A computerized payroll, finance, inventory, and invoicing system was developed at the R R Nagar plant as early as 1985.

The same model was replicated at the Jayanthipuram, Alathiyur, and Mathodu plants as they became operational. In addition to having a self-contained IT infrastructure, each plant also boasted an internal IT organization to support the deployed infrastructure. However, the growth in IT investments across all plants was not proving to be sufficient. (**Exhibit 7** details the IT organizational structure, **Exhibit 8** provides details of IT infrastructure at Madras Cements, and **Exhibit 9** depicts profiles of users at Madras Cements.)

By the mid-'90s, the IT infrastructure and the supporting organization had grown quite differently at all the four plants. Each plant was essentially a closed system with its own set of systems, platforms, and processes. There was no sharing of information between the plants. As a result of the disparate processes, the corporate headquarters received different metrics from each plant. Not only were the metrics different, each plant measured different things as well, making comparison and benchmarking very difficult. For example: the Alathiyur plant would send a "Combined Parameters at a Glance" report but the other plants would not. And to further complicate the situation, each plant reported its metrics at different times during the month. The fact that all the reports were consolidated manually at the corporate level not only delayed the final reporting but increased the propensity of introducing human error.

Missing the Right Reports

Corporate headquarters was spending a significant amount of time and resources to standardize the reports that it prepared for management. In addition, each report that was requested by management necessitated a couple of days of work by analysts to "massage" the available data in the format requested. Not only were the requested reports not available in real-time, even their reliability was questionable, as the underlying information was not standard across plants and the significant "massaging" of data made it very difficult to reconcile information from different plants. Moreover, there were always differences between the corporate analysts and the plant personnel over the accuracy of the figures in the reports. So not only was Madras Cements expending resources in creating basic management reports, the management was making decisions based on inadequate, incomplete, and inaccurate information.

As the management struggled to reduce costs, it was realizing that its existing reporting controls were inadequate to manage its scale and growth.

> Reports were prepared by the 10th of the month, the different manufacturing plants held discussions by the 12th, and senior management met to discuss the reports around 20th of the following month. This time line was too stretched to react to the market.
> We, in the past, never looked at efficiency. Our margins increased primarily by increasing prices.
>
> \- A.V. Dharmakrishnan, CFO

It was very clear to the senior management that this situation could not continue for long, and significant efforts were required to standardize processes across plants and establish the technology infrastructure needed to enable management to monitor and control costs effectively.

First Implementation: An Unsuccessful Attempt

Ramco Systems, a sister company of Madras Cements, had started operations in 1989. It specialized in the development and deployment of enterprise resource planning (ERP) solutions. By mid-1990, it had just finished the development of its e.Applications ERP suite and was in search of a client where it could showcase its new solution package. At the same time, sister company Madras Cements was experiencing difficulties in its day-to-day operations due to the differing systems and processes at its plants. The lack of a connecting infrastructure between locations had forced each plant to develop its own independent identity, which in turn was also different than corporate headquarters. Lack of standardization of systems and processes was hampering the management in making accurate operational decisions.

Ramco Systems was aware of the difficulties at Madras Cements and offered to assist Madras Cements solve its problem by deploying the recently completed e.Applications ERP suite as a solution to integrate the different IT islands at each plant and corporate headquarters. This came during the boom time for ERP and it was being promoted as the most effective IT solution for any company. However, there also were innumerable examples of failed ERP implementation. In fact, some companies even went out of business as a result of failed ERP implementations. Employees of Madras Cements were aware of the difficulty in successfully implementing ERP and were hesitant to embrace the proposed ERP system. Even among the senior management there was reluctance to sign off on the ERP implementation exercise. At that point in time, the ERP system was still considered the exclusive domain of the IT functional division, and hence the risks and the business consequences of an unsuccessful ERP implementation were not thoroughly analyzed. As a result Madras Cements did not view the ERP vendor selection process to have the same importance as, say, selecting a power provider. This resulted in the vendor selection process not being as well formed as it could have been. Nevertheless, a decision was finally made to allow Ramco Systems to deploy its e.Applications 1.0 suite at Madras Cements. One of the primary drivers for Madras Cements management to go ahead with this decision was to assist its sister concern in achieving critical mass as a significant competitor in the ERP space.

The proposed ERP system, e.Applications 1.0, while being a completely developed system, had never been tested in a production environment and also not customized. As a result it was not yet mature enough to produce immediate visible benefits to Madras Cements. Users were not fully trained in the use of ERP data for decision making. They were not aware of the importance and the impact of the integrated, online nature of the information available through ERP. They even felt that they would be better off without ERP.

Compounding the issue was the approach taken by the project team. Senior management was not involved in the project, and the entire exercise was viewed more as an IT upgrade rather than something affecting the entire business operations of the company. As a result, mapping the existing processes and streamlining the process flows were not even considered to be within the scope of the project. The results of the implementation were below what management expected. While the implementation of e.Applications for some functions at some plant locations was successful, it did not provide any overall business benefits.

The costs and the resources (approximately Rs. 10 million or $244,000 for e.Application 1.0) spent in the ERP implementation caused the senior management of Madras Cements to sit up and take notice. The management saw that the ERP system was not functioning as expected and the information being collected and received from different plants was not standardized, which in turn was hampering management's ability to make accurate operational decisions. One key indicator was that the collection and collation of power consumption data at each plant took anywhere from a week to a month. Given the heavy power consumption needs of the cement industry this delay was hampering the ability of the senior management to effectively control one of the biggest cost components. After the unsuccessful IT implementation, senior management members felt that they were going to a worse state than they were before. So they pondered a central question:

> Are ERP systems inherently not as productive as hyped or was Madras Cements' bad experience due to factors other than the ERP system itself?

Rethinking the Approach

The senior management at Madras Cements started to rethink its approach to managing technology. Executive managers took external inputs and internal suggestions and met to discuss how they could leverage collective learning from the unsuccessful attempt earlier. Considering that the recent ERP implementation was fresh on everyone's minds, it was the most discussed issue. While the entire executive management agreed that ERP did seem to provide the best solution for the current operational ineffectiveness, managers were skeptical about how to avoid repeating the same mistakes that plagued the first ERP implementation. There were also concerns about who among the executive management would be best suited to champion the ERP implementation.

Social Transformation

With fresh knowledge and insights acquired from the recently concluded executive program, executive management embarked on a series of initiatives to re-implement ERP successfully. The management recognized that the primary hurdle, during the earlier implementation, was people's reluctance to give up the current culture and practices of information use. Hence, the initial challenge for Madras Cements was to think of the practical ways to transform this mindset.

First and foremost, it was felt that senior management should not only be significantly involved but in fact be driving this project. As a result, Dharmakrishnan, the CFO and one of the participants in the Michigan executive program, took the responsibility of leading the ERP implementation. Secondly, senior management realized that the goal of the ERP implementation project was not just a successful deployment, but that the real purpose was that the concerned end-users should utilize the information that is made available by the ERP system for tangible business results. Beyond deployment, IT implementation was less about technology and more about people. Analysis of the failed ERP implementation effort by Madras Cements revealed that the main cause of the failure was putting more focus in technology and less focus on the people and the needs of the executives. Technology obviously resides with the vendor, but the processes are intrinsic to the company. Hence, a realization sunk in that Madras Cements would have to take a bigger responsibility for the ERP implementation than that taken by the ERP vendor.

Thirdly, the team realized that assuming individuals will use technology or data because it is available is a misconception. Unless management devotes the required resources for training, the implementation will not be successful. Madras Cements also recognized the need to build a culture of information transparency and knowledge-sharing across the plants. Dharmakrishnan noticed that the plants were disconnected, in the sense that information was not shared among them, and managers were still running their businesses with the same old regional fiefdoms and inadequate ways of satisfying customers. He wanted to take the best capabilities they had and leverage them in all operations across the country. In order to accomplish the desired culture, he felt that he needed to encourage a system wherein people were adept at exchanging ideas, processes, and systems between plants, and wherein people were absolutely free of the "not-invented-here" syndrome -- part of the existing social mindset.

Finally, in any organization employees question and challenge major changes all the time. They don't accept things at face value. As a result, Dharmakrishnan felt that it was his responsibility to convince employees why transformation was necessary. For a company to truly establish its presence across India, employees had to change the way they thought and acted, taking on progressively more responsibilities and initiatives.

Fresh Start: New Mindset and Managerial Orientation

In order to create and encourage a culture of transparency, Dharmakrishnan initiated a process-oriented approach: map out existing processes and create a set of "to be" flexible and transparent processes, which would be standard across all the plants. In this way he was trying to adopt a companywide total-quality management system. As outlined earlier, Dharmakrishnan was now convinced that if the ERP implementation was to be successful then it would need to be process-driven and produce operational results, which in turn would lead to improved operational efficiency. The bigger challenge was to take this realization and put it into practice. But he was sure of the fact that it was this transformation (of these business processes into a flexible and predictable system) that would provide real-time visibility and thus enable cost competitiveness, which was the primary desired outcome of the implementation. Providing a clear process map and technical description of business processes facilitates clear communications among the various stakeholders.

His first few experiments exposed the quality of existing processes and decision-making, and showed how inefficient they were. But the group didn't want any of the new one-company vision. They saw it as a threat. Yet Dharmakrishnan didn't expect changes to happen overnight. Exposing employees slowly to new ideas helped Dharmakrishnan change the orientation and mindset of managers to accept a new competitive reality and corresponding managerial practice. For this, he adopted a phased approach (as against a "big-bang approach"). Phased implementation allows gradual increase of the transparency in business operations at and to various levels of management. He also started encouraging users to focus on understanding which sets of information, available as a result of a successful ERP implementation, would enable Madras Cements to derive significant cost, quality, and cycle-time improvements. This was consistent with the message given in the Michigan executive program.

Finally, the implementation team consisted of cross-functional members -- a team constituted of analysts from corporate headquarters, IT personnel, functional experts, and implementation experts from Ramco Systems under the overall guidance of the general manager of IT. Additional resources required from the different functional areas and from the respective plants were brought in on an as-needed basis.

New Implementation: Second Chance

In the analysis of the last implementation effort that Ramco Systems conducted after the e.Applications 1.0 did not perform to expectations, it was recognized that key benefits of any ERP package could be realized only if it were flexible enough to fully incorporate the processes of the client organization. This was the key takeaway by Ramco Systems from the unsuccessful implementation of e.Applications 1.0 at Madras Cements. Ramco Systems tried to incorporate the learning into its ERP product. When Dharmakrishnan announced his intention to lead the new ERP implementation efforts at Madras Cements, Ramco Systems was ready to assist its sister concern with a revised version of its e.Applications suite.

Madras Cements' management decided to give another chance to Ramco Systems for the second ERP implementation effort. It was imperative that Ramco Systems demonstrate a successful implementation at Madras Cements in order to be able to successfully compete in the ERP market, and Madras Cements decided to help its sister company establish its credentials. Thus, the e.Applications 3.0 system by Ramco Systems was chosen as the ERP package for Madras Cements.

Phased Approach

A phased approach calls for rolling out an ERP package at one location first and one module at a time. The benefit of this approach is that if an error is detected at one location it can be fixed before the whole ERP package is deployed at all locations. In addition, with a phased approach the resource requirements are lower because the implementation is spread over a longer time frame. Further, the experience gained from implementing one module or location can be applied to the next module or location being readied for deployment. A phased approach presents an opportunity for progressive implementation of ERP with minimum risks. The main drawback of this approach is that it essentially requires a company to run the legacy system and for the new ERP system in parallel till all the sites have the ERP package deployed.

Dharmakrishnan was aware that the prudent approach would be to first implement e.Applications 3.0 at one plant, work out the kinks in the implementation, and then copy that implementation to another plant. Because all the plants were operating independently there was no benefit to be gained in implementing e.Applications at all plants simultaneously. Hence, Madras Cements decided to follow the phased approach for ERP implementation. This would allow for gradual adoption at the user level and training on one module at a time as more beneficial than training in all. User acceptance and willingness to effectively use the system, as outlined earlier, was something Madras Cements really wanted to address during this course of implementation.

Process Redesign and Training

The first task of the core team was to map out existing processes across nine different functional areas/ modules. The processes were different for each plant. These separate processes were taken as input and used to create a set of "to be" processes, which would be standard across all the plants. Dharmakrishnan spent six months with the core team discussing the overall direction of the business process. Subsequently the team took charge of mapping existing processes, identifying the opportunities for improving processes, and converting them to "to be" processes. The team had to also account for the fact that the communications infrastructure to link the plant sites was not robust enough to allow for every transaction to be executed in real-time. (**Exhibit 10** gives an overall view of the communications infrastructure available.) Team members had to design the processes in such a way that approximately 95% of processing happened locally -- within the local IT infrastructure of each plant. The team took about a month per module to map the existing processes and convert them to the "to be" processes. The testing of the "to be" processes was also accomplished within this month.

In order to prepare the rest of the company for the coming ERP implementation, Dharmakrishnan initiated a recurring program to train future users on the ERP system. The training was primarily provided by Ramco Systems and covered a basic overview of ERP as well as details about the impact of ERP on each of the nine modules. Every employee of Madras Cements underwent the basic training. The training, apart from educating the employees, also served to allay any latent fears about job loss among the employees.

Once all the processes were in place after being fully tested, Ramco Systems started deploying the e.Applications in November 1999 at R R Nagar, the first and the oldest plant. The goal was to deploy the complete system at one facility first, stabilize the release by working out all kinks and bugs, and then take the stable release to the next manufacturing facility. A month after successful deployment of e.Applications at R R Nagar, JPM went online with the e.Applications suite. In the next few months all of the plants and the corporate headquarters were successfully running and utilizing e.Applications 3.0.

Customizing Reports: Information Pull vs. Push

When the ERP suite was implemented, there were built-in reports incorporated in the system. However, many users felt that these canned reports were inadequate for their analysis and decision-making process. Since the real success of ERP was to be found not simply in implementation, but rather its use to improve operational efficiency and decision-making, Dharmakrishnan emphasized customization of the reports required by the users through in-house efforts. He formed a team of IT analysts within Madras Cements headed by Mr G. Murugesan for this purpose. The team was tasked to do the following:

- Develop user-friendly reports that could be understood by every employee
- Interface all non-ERP applications such as weigh bridge, attendance recording system, fixed deposits, etc., with the ERP system
- Capture process data and incorporate that in the ERP system in real time
- Train the users in proper and efficient use of such reports
- Study and improve security of data

The team developed a set of MIS reports under the overall guidance of Dharmakrishnan and the IT staff. Users were also trained to utilize such reports for improving their effectiveness. (Exhibit 11 provides screen shots of some these reports.)

Business Benefits: The Essence of Transformation

When Dharmakrishnan first initiated the entire ERP exercise he did not even have a computer at his desk and was not using any computerized processes to track the business. But as he started devoting more time to the ERP project (almost 70% of his time in 2003), he began to internalize all the benefits that a successful usage of the ERP system would bring to Madras Cements.

One interesting example of a benefit from the ERP system was how ERP generated data that was used to both reduce inventory and optimize logistics. Traditionally, Madras Cements had maintained 40 warehouses loaded with various cement products to supply its distributors. With the new analysis based on ERP data, Dharmakrishnan concluded that Madras Cements needed only nine warehouses to maintain its standards of delivery. However, this analysis was opposed by some of the old guard. They voiced their opposition to the plan to reduce the warehouses by referring to the state regulation that prevented delivery trucks from entering and leaving major markets during business hours. They argued that all the warehouses were needed to ensure deliveries. Instead of having to back down, this gave Dharmakrishnan yet another business process to improve. He went back to his data to see if anything could be done about the way delivery truck operation was managed. While doing the analysis he came to realize that the way delivery trucks were being dispatched essentially forced them to spend most of the day waiting at rest stops. After consulting his production planners, Dharmakrishnan and his team were able to revise the scheduled pickup and drop-off times for the delivery trucks, improving the usage of the fleet by 50% and still demonstrating that Madras Cements needed only nine warehouses. Reduction in the number of warehouses gave Madras Cements a cost saving between $9,756,097 and $12,195,121. This enabled the company to be the only one in its field to make profits during the years 2003 and 2004, when all its competitors were taking huge losses. Following this episode, the opposition to the ERP system weakened and Dharmakrishnan was quickly able to bring about the following additional benefits:

Customer satisfaction: As ERP provided a uniform system across plants and corporate headquarters, Madras Cements could follow up on customer orders to ensure higher customer satisfaction. In fact, after ERP was implemented, orders were shipped in less than 24 hours of being received.

Plant efficiencies: Now that everything was automated and standardized, Madras Cements could benchmark one plant's performance against other plants, thereby ensuring better efficiencies.

Improved management transparency: With the implementation of "dashboard" aggregating of critical information for Madras Cements' management, the management had a real-time understanding about every aspect of Madras Cements' business. Now instead of waiting half a day every day to receive and review reports, managers could act upon the real-time information they had and improve the business.

Having experienced first-hand the power of technology, Dharmakrishnan wanted to ensure that the entire company benefited from IT. He had IT personnel attend business meetings as observers to collect information about what was discussed. He then advocated that the personnel use that knowledge to create/modify reports that would better serve the needs of these users and the business. Although there was some resistance from the business groups when they were asked to change how they operated, the resistance quickly dissipated once the business groups realized that as a result of IT they were able to recognize significant cost savings. As a result of these changes Madras Cements evolved to have the best-cost structure in the cement industry in India. Successful deployment of e-Applications 3.0 across all the plants and the corporate headquarters resulted in significant cost savings and efficiency improvements, identified as follows:

Production	• Overall operations consistency was achieved, and productivity was enhanced from 5 to 10 tons per hour. This implied recurring annual savings of about $1.8 million. • The power generator utilization factor was increased by 10% and power (electricity) consumption was reduced by 10 units per ton by continuously monitoring factory operations using real-time data in ERP. This indicated recurring annual savings of about $4 million. • Expected cement bag weight was achieved for 99% of production, resulting in recurring annual savings of about $2 million. • Weight variation in the cement bags came down from 40% to 2%, resulting in net savings of $2 million annually. • On average, variable costs decreased by $6 per ton.
Materials	• Better prices were realized from the vendor by comparing the unit prices, availing goods discounts, and better credit periods. • Inventory levels were reduced by monitoring materials received but not by materials consumed within the committed time. This resulted in recurring annual savings of approximately $0.45 million.
Management Accounting	• Variable costs were analyzed on daily basis for each process center. • Fuel efficiency was analyzed with caloric value and the market price of the items, arriving at an economical fuel mix.
Finance	• Trial balances of all the factories were analyzed with greater detail. • All administrative overheads were reduced without affecting the effective operations. Reduction was achieved mainly by process redesign. For example: With TT charges (bank charges for non-local transactions), all major payments were now made locally by negotiating with the excise/sales tax/electricity authorities rather than transferring the funds to factories. Similar reduction of administrative expenses resulted in a recurring annual savings of about $0.45 million. • 100% adoption was achieved for the costing system, which updated the P&L for the entire firm in real time upon entry of a transaction.

Sales	• With the close follow-up of all pending orders, orders could be executed within 24 hours. This led to increased customer satisfaction. • Transporters' freight was analyzed on daily basis. Based on this, logistics were derived. Stock transfers to depots were handled without any re-handling process. • Analysis of ERP data led to closing down more than 90% of stock points, which enabled the company to save on stock holding, transportation, and re-handling. This resulted in a recurring annual savings of about $3.6 million.
Overall	• Performance was analyzed on a mine, equipment, and shift-wise basis. Based on this analysis, about 60% of heavy equipment was withdrawn from the operations due to poor performance or underutilization. • The number of shifts was brought down from three to two. • Re-handling of materials was brought down to almost negligible from an earlier rate of $0.40 per metric ton. This resulted in a recurring annual savings of about $0.7 million. The company realized $22 million in cost savings overall as a result of successful ERP implementation, with more to come.

Web-based Centralized ERP

Having tasted success in implementation of ERP Version 3.0, the company continued to use the application till 2008, when a need was felt for improvements in the system. That is when the company decided to upgrade the ERP to the version 4.2, which was a web-based and centralized system. The previous system was decentralized and desktop-based. It took just six months to install the new system and it was smooth, thanks to the experience the team had gained in the previous implementation.

The project team constituted for this purpose interacted with the users extensively during pre-implementation, implementation, and post-implementation stages. The team put across to the ERP vendor what the users actually wanted, thereby bridging the gap between the expectations of the users and the vendor. This was the key that facilitated smooth implementation. Madras Cements received tremendous support and assistance from the vendor Ramco Systems, which also helped take care of this aspect.

To ensure that the fruits of the ERP system reached all the levels, a special report was added to the MIS system, which showed the access pattern of various users and their use of ERP and MIS reports. Top management, including the chairman, monitored this data and gentle reminders were given to those who were not utilizing the power of the ERP system properly. (A sample letter from the chairman is enclosed as Exhibit 12.) One-to-one follow-up sessions were also held with key users to get the field-level feedback to sort out implementation problems.

Besides providing a better user interface, the new system brought the benefits of improved control due to centralization. There was now more consistency in masters, and across the organization comparisons were made easy by using this system. In addition, it was interfaced with a strong master data management system that ensured non-duplication of master items, an essential ingredient of centralized IT setup.

Strengthening ERP

ERP became the backbone of Madras Cements' information infrastructure and the workhorse of all major business processes of the organization. The IT team, headed by N. Varadarajan, senior general manager-IT, visualized that Madras Cements could obtain a multiplier effect if it built some innovative systems that could supplement and complement ERP. These IT tools and technologies were also intended to help the company to monitor, streamline, and secure its processes. That was the beginning of a project called Beyond ERP, which saw innovative systems interfacing with ERP and maximizing the benefits. Following are some of the prominent systems under this project:

- **Enterprise mobile computing:** The company developed an in-house enterprise mobile computing (EMC) system that facilitated business processes like order booking, invoicing, and pack-slipping through PDAs, by integrating those devices with the organization's ERP system. The application proved to be quite useful for mobile employees like sales professionals. A complete sales order booking cycle could be facilitated through this; more than 80% of orders were booked through these PDAs.

This challenging project was entirely carried out by an in-house team headed by P. Nagendran, assistant manager-IT, and guided by Murugesan, DGM-IT. To illustrate the impact of this project, the following table gives the status before and after the EMC implementation.

Process	Before EMC	After EMC	Impact
Pre-order information	The salesperson has to check with the sales office or maintain his or her own paper database	At any point of time a salesperson can use a PDA interactive information system to know about a pending order, credit limit, etc., pertaining to a customer	Information on demand and better customer satisfaction
Order booking	The salesperson calls someone in the regional office, who books into a desktop ERP periodically	The salesperson books through a PDA at any time	Empowerment of salesperson; customer satisfaction improves because of 24/7 order booking
Confirmation of booked order	The salesperson has to contact the regional sales office by phone	The salesperson gets the confirmation within one minute in a PDA	Sales productivity rises with fewer unnecessary phone calls
Order status	Not available except through checking with the factory dispatch section	Constantly updated automatically in the PDA whenever status changes	Sales productivity rises
Depot invoices	There is a requirement for hardware and complicated MPLS lines or some other mode of connectivity	The simple mobile GSM / GPRS connectivity is sufficient	Costs saved, operations simplified
Depot invoices	Portability is the problem at railheads	Portability is not a problem	Online invoicing at railhead possible
Wagon dispatches	Invoices updated on the next day due to problems in connectivity at railhead	Invoices updated online immediately upon dispatch	Stock positions updated online; better decision-making.

With so many benefits, the project received exceptional response from salespeople. Plans were in the works for incorporating more and more processes like approvals, credit limit revision, and daily call reports into this PDA system.

- **Mobile MIS reports:** Through an MIS (management information system) customized for mobiles, top management could access these reports through their mobiles, 24/7, anywhere around the globe.

- **Leveraging the power of SMS:** Madras Cements made use of an interactive SMS platform to facilitate customer queries related to the status of their orders/checks/last three payments. It also used push-SMS system to send data updates automatically with regard to different processes' status to the customers and top management officials. For example, when a cement truck left the factory, an SMS went to the concerned salesperson and customer indicating time of dispatch, truck number, quantity, etc.
- **Compliance software on top of the ERP:** This software generated reminders/alerts regarding statutory and operational compliances. The application would repeatedly issue warnings from ERP data to officials related to necessary process/statutory compliances that needed to be adhered to. In case of failures, it warned higher officials too.
- **Dealing with security at all levels:** For ensuring the security of information in the ERP exchanged over network, Madras Cements adopted a number of proactive measures, like:
    - An online ERP rights management system
    - Firewalls (both hardware and software)
    - VPN for all connections through the Internet
    - Online central antivirus system
    - Centralized AVS updating and patch management
    - Information security policy
    - Proximity and biometrics security solutions
    - Disaster recovery center
- **Tracking the IT assets:** An infrastructure management system stored the data on the vast IT infrastructure, comprising desktops, laptops, PDAs, printers, servers, etc. At any point of time, by click of mouse, it could be seen which location had the particular type of asset. This system was very useful in planning replacements, handling requests for new hardware, etc.
- **Interactive voice response (IVR) system:** An IVR system was developed in-house. The system was a technology in which a telephone could be used to interact with an organization or its database management system, like ERP, to acquire information from or enter data into the database.
- **For the sake of ecology:** Madras Cements also adopted "paperless office" as an important organizational strategy. To accomplish this goal it developed an online e-approval system through a workflow process integrated with ERP. Instead of paper documents, only electronic documents flowed through various reviewers and approvers. Approvals could be done through computers and mobiles. This practice saved a lot of paper and also helped to track important documents at various stages.

The Team Behind the Success Story

Madras Cements had a strong IT department of about 60 people who were the force behind the ERP implementation and the innovative systems detailed above. All were professionals having an average age of 25 to 30 years. The team was headed by N. Varadarajan, senior general manager-IT, with the able and helpful deputy general managers Murugesan and Muthukrishnan supervising a group of motivated young professionals. (The organization of the IT department is shown in Exhibit 10.) Madras Cements appeared to have been one of the very few manufacturing companies that could boast of an in-house IT team of this size and capability. It had a set of developers who focused on emerging technologies for industries similar to it; another group provided technical and operational support to the ERP users. An in-house data center hosted all the servers and provided support around the clock. The IT team also included information security personnel, because Madras Cements believed that security of data was very important for any IT projects. Because of all this in-house talent, Madras Cements felt itself to be in a position to address any challenge that might arise at any point.

Having installed a good ERP itself is only the first step; the most important thing is running ERP and the allied IT processes effectively and efficiently. The IT department set for itself challenging goals in efficient IT operations and IT cost reduction. The following table shows the IT cost reduction status near the end of the first decade of the 21st century through optimizing inputs and service providers and by better negotiations.

Area	Annual Cost in $Thousands	Target for Reduction	Status
Connectivity	344	50%	Achieved
AMC	98	12%	Achieved
ERP Support	375	40%	Achieved
DR Site	94	44%	Under Progress

Madras Cements began a project to reduce replacement cost of hardware and licensing of software for the coming years. It planned to use technological advancements like cloud computing and virtualization for this exercise.

Conclusion

Before the ERP implementation, each manufacturing facility manually entered its power consumption in an Excel spreadsheet. The corporate office received these spreadsheets from each facility after the month, which hampered the management's ability to react to one of the key cost components affecting the cement business. After the implementation, the power usage information was available in real-time via a browser interface to the senior management. Due to better integration of dealer performance data, Madras Cements was able to offer end-of-year incentives to its dealers, which helped in strengthening its relationships with key dealers. Also because Madras Cements was able to benchmark its performance at a granular level, it was able to experiment with its production processes by switching to local Pet Coke when the prices of coal shot up by 6%. All the benefits that Madras Cements accrued from the successful ERP implementation were on the cost side. The questions for the future facing Dharmakrishnan were:

Could Madras Cements leverage its IT infrastructure to influence the revenue side by expanding to other Indian markets? Diversifying into ready-mix cement? Influencing the entire construction industry by introducing "just-in-time" process?

Exhibits

Exhibit 1

Ramco Group of Companies

Exhibit 2

Plant Locations

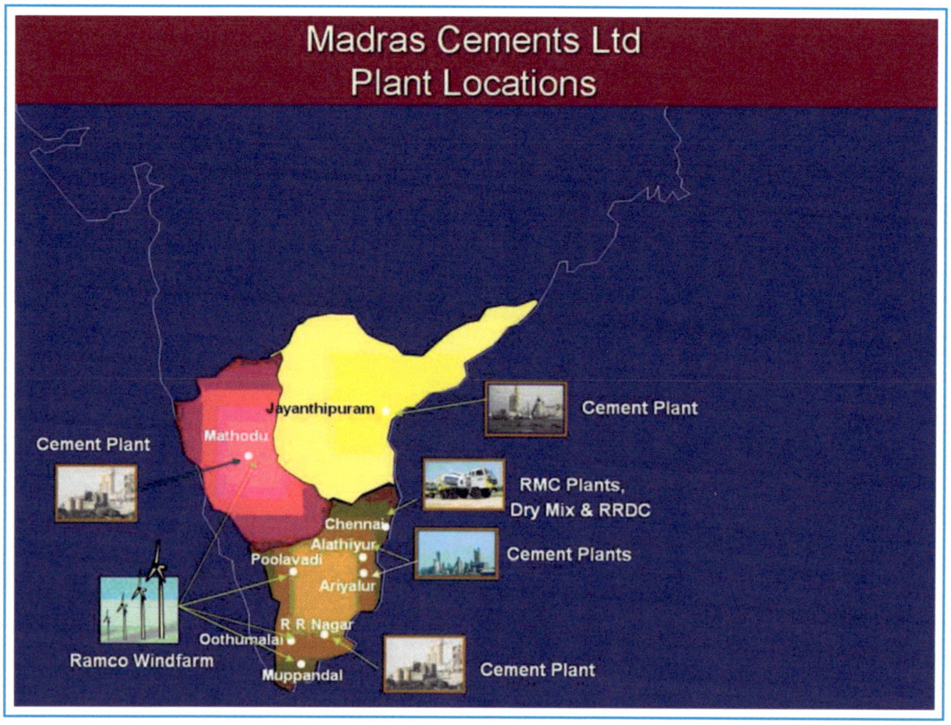

Exhibit 3

Cement Manufacturing Process

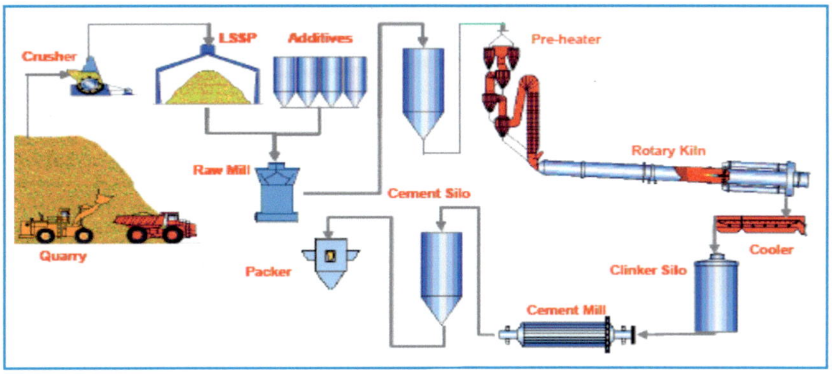

Exhibit 4

Historical Cement Prices in MCL's Key Markets (in Rupees)

STATE	TAMILNADU					
Year	1998-99	1999-00	2000-01	2001-02	2002-03	2003-04
Month						
April	141	142	127	173	150	148
May	147	143	126	166	150	148
June	155	143	133	157	145	148
July	171	158	152	159	140	143
August	171	172	161	135	125	138
September	166	169	163	127	120	133
October	162	160	167	151	115	
November	153	148	174	157	128	
December	150	136	175	166	128	
January	143	123	181	168	148	
February	137	121	182	153	148	
March	139	132	177	145	148	
Avg. Price for the year	**153**	**146**	**160**	**155**	**137**	**140**

STATE	KERALA					
Year	1998-99	1999-00	2000-01	2001-02	2002-03	2003-04
Month						
April	148	155	147	192	154	164
May	159	156	143	186	148	168
June	166	156	148	178	142	165
July	183	171	170	172	135	160
August	184	185	181	147	135	142
September	171	185	180	146	125	136
October	159	181	183	165	120	
November	149	172	186	166	140	
December	145	154	187	175	137	
January	142	135	195	175	157	
February	138	132	196	165	160	
March	144	149	196	157	162	
Avg. Price for the year	**157**	**161**	**176**	**169**	**143**	**151**

STATE	KARNATAKA					
Year	1998-99	1999-00	2000-01	2001-02	2002-03	2003-04
Month						
April	128	117	116	146	115	140
May	134	120	113	142	114	146
June	140	121	113	133	116	142
July	141	123	126	133	120	132
August	135	132	136	131	122	122
September	125	126	139	118	118	116
October	121	120	142	130	125	
November	118	116	142	132	130	
December	120	110	147	139	133	
January	119	106	152	148	129	
February	115	107	152	137	132	
March	115	115	149	123	136	
Avg. Price for the year	**126**	**118**	**136**	**134**	**124**	**131**

STATE	ANDHRA PRADESH					
Year	1998-99	1999-00	2000-01	2001-02	2002-03	2003-04
Month						
April	114	102	109	136	105	122
May	126	104	98	138	103	120
June	122	110	99	136	98	110
July	130	119	121	133	92	104
August	118	125	134	125	90	95
September	104	114	139	106	100	92
October	96	107	144	128	125	
November	96	100	145	125	133	
December	100	97	147	126	123	
January	96	96	150	128	115	
February	94	99	152	120	112	
March	100	116	143	112	120	
Avg. Price for the year	**108**	**107**	**132**	**126**	**110**	**110**

Exhibit 5

Cement Demands Across india

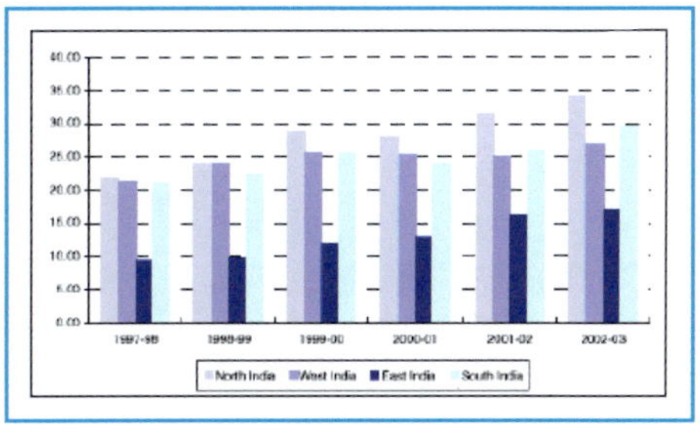

Exhibit 6

Market Share

(per cent)	1993-94	1994-95	1995-96	1996-97	1997-98	1998-99	1999-2000	2000-01	2001-02	2002-03
ACC	14.3	14.5	13.8	12.2	13.2	11.6	11.2	11.4	12.2	12.8
Grasim	4.3	4.5	5.2	6.4	6.0	5.3	10.0	10.9	10.3	10.9
Larsen and Toubro	4.4	5.1	6.0	7.0	8.8	9.9	11.6	11.5	11.1	10.5
Gujarat Ambuja	2.8	3.0	4.0	5.8	5.1	7.6	7.2	8.4	8.7	9.5
India Cements	4.5	4.1	4.1	4.3	6.7	6.7	7.9	7.6	6.4	5.4
Century Textiles	5.5	5.2	4.4	5.5	5.4	5.0	5.2	5.5	5.0	4.8
Birla Corp.	5.9	5.0	5.3	4.5	4.7	3.6	3.9	4.1	4.0	4.1
Jaiprakash Inds	4.2	4.1	3.4	3.4	3.1	3.6	3.9	3.9	3.9	3.8
Lafarge	-	-	-	-	-	-	0.9	2.3	3.8	3.4
Madras Cements	2.5	2.6	2.9	3.1	2.7	3.3	2.9	2.9	3.1	3.1
JK Synthetics	3.0	3.0	2.7	2.3	2.5	2.2	2.2	2.1	2.3	2.7
Shree Cement	1.5	1.6	1.4	2.0	1.3	2.5	2.4	2.6	2.4	2.5
[illegible]	2.5	2.6	2.5	2.7	2.0	2.4	2.3	2.0	2.3	[illegible]
Zuari Inds	0.9	0.6	0.9	0.8	0.8	0.7	1.6	1.6	1.6	2.1
JK Corp.	1.0	1.0	1.4	1.9	1.7	2.0	2.0	1.9	1.9	2.0
Binani Cement	-	-	-	0.7	0.6	1.7	1.8	2.0	2.0	2.6

Exhibit 7

MCL IT's Organization

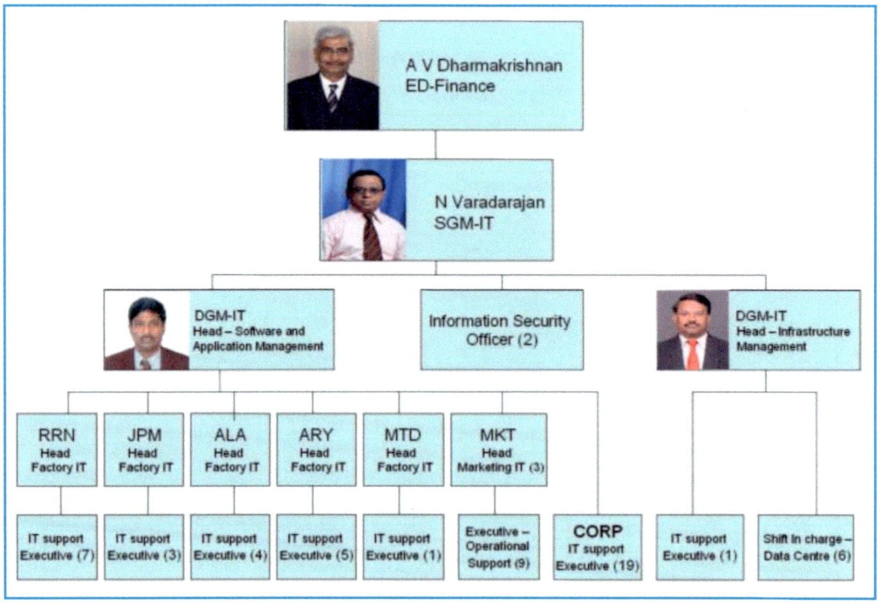

Exhibit 8A

IT Infrastructure: Machines

Server Type	Description	Function
ERP Server	P-II * 4 / 2GB RAM / 64GB HDD	All Databases related to ERP Applications are maintained.
PDC-NT Server	P-III / 128MB RAM / 4GB HDD	Primary Domain Controller. All users logins are validated
Terminal Server	P-II * 2 / 1GB RAM / 18GB HDD	Remote users access the ERP Server through this to get better throughput on WAN using the features available in Windows-2000 system like Terminal Server, Terminal Client Access software.
Mail Server	P-III / 128MB RAM / 4GB HDD	All our Internal & External mails are routed through this Mail (Messaging) Server using software like MS-Exchange & MS-Outlook.

Exhibit 8B

IT Infrastructure: Number of Workstations

Module	RRN	JPM	ALA	COR	MKT
Finance (GL & AP)	11	6	12	9	-
Logistics (POM & IMS)	8	7	10	2	-
Maintenance (EQP & MO)	10	12	16	-	-
HRM (HR & Payroll)	3	4	4	5	-
Sales (SOM & SHP)	4	3	5	10	-
Mines (ORE)	8	3	3	3	-
Production (CPP)	6	1	3	-	-
OCM	3	2	3	-	-
Marketing	-	-	-	-	52
General	50	35	61	22	-
Total	103	73	117	51	52
				Total	396

Exhibit 9

Number of Users

Module	RRN	JPM	ALA	CORP	Total
Finance (GL & AP)	11	10	12	9	**42**
Logistics (POM & IMS)	8	11	10	2	**31**
Maintenance (EQP & MO)	47	43	53		**143**
HRM (HR & Payroll)	8	5	8	5	**26**
Sales (SOM & SHP)	4	3	5	10	**22**
Mines (ORE)	33	17	17	4	**71**
Production (CPP)	13	9	8		**30**
Marketing	15	9	15	3	**42**
Total-ERP Applications	**139**	**107**	**128**	**33**	**407**
Non-ERP (OCM,Real-time systems, Stenos etc.)	**50**	**35**	**61**	**22**	**168**
Total	**189**	**142**	**189**	**55**	**575**

Exhibit 10

MCL's Network Infrastructure

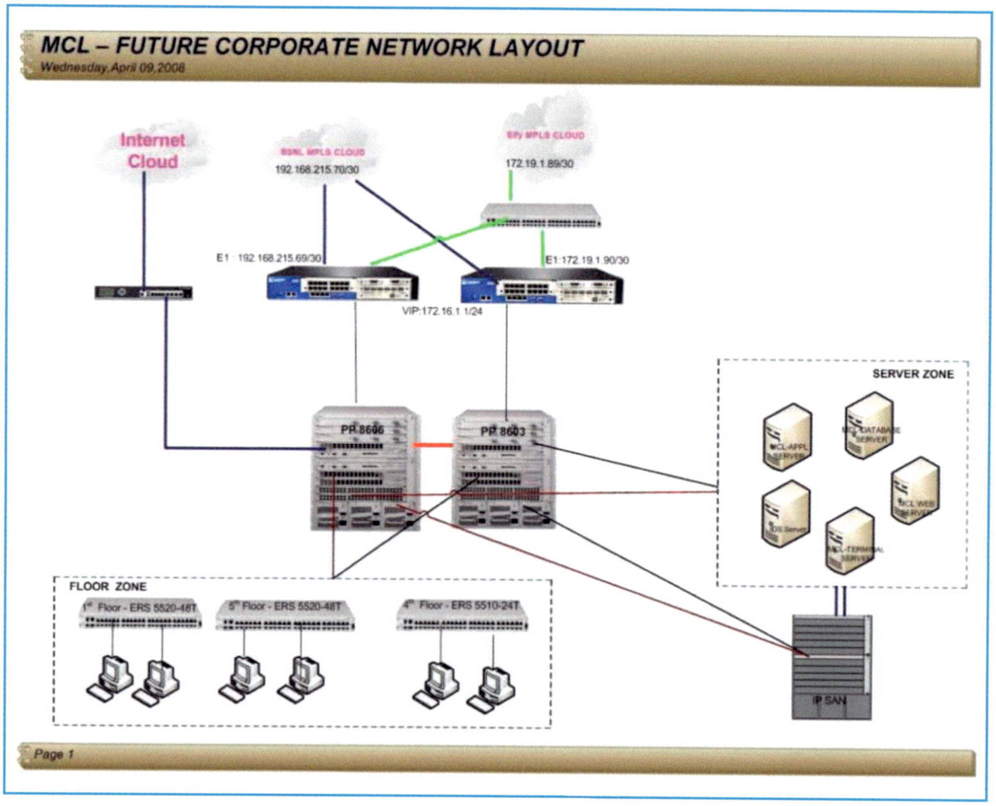

Exhibit 11

Screenshot of Sample Reports

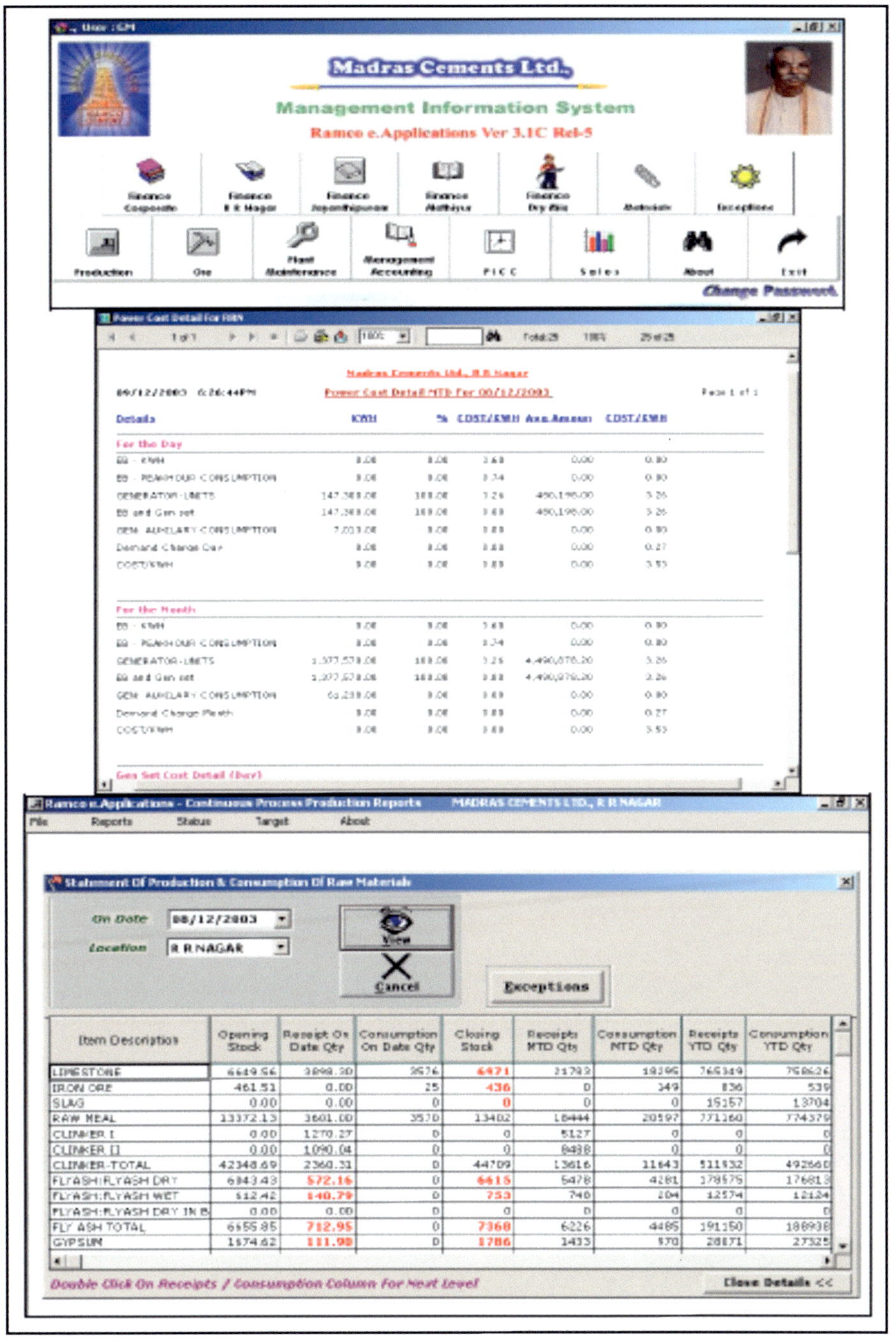

Notes

Notes

Notes

Made in the USA
Las Vegas, NV
07 November 2026

57672683R00017